THE OPEN UNIVERSITY

Mathematics: A Third Level Course

Numerical Computation   Unit 1

Introduction to Numerical Methods

The Open University Press

Course Team

Chairman:	W. Y. Arms
Central Members:	P. A. Blachford
	T. M. Bromilow
	F. C. Holroyd
	P. G. Thomas
Staff Tutors:	M. Kennedy
	J. E. Phythian
B.B.C.:	J. A. Richmond
	R. I. Clamp
Student Computing Service:	F. L. Irvine
	M. B. Haywood
Consultants:	L. Fox (University of Oxford)
	H. P. Williams (University of Sussex)

The Open University Press, Walton Hall, Milton Keynes.

First published 1975.

Produced in Great Britain by
Technical Filmsetters Europe Limited, 76 Great Bridgewater Street, Manchester M1 5JY.

ISBN 0 335 05650 4

This text forms part of the correspondence element of an Open University Third Level Course. The complete list of units in the course is given at the end of this text.

For general availability of supporting material referred to in this text, please write to the Director of Marketing, The Open University, P.O. Box 81, Walton Hall, Milton Keynes, MK7 6AT.

Further information on Open University courses may be obtained from The Admissions Office, The Open University, P.O. Box 48, Walton Hall, Milton Keynes, MK7 6AB.

1.1

Contents

	Page
Guide to the Course	4
Study Guide for Units 1 to 4	6
Objectives for Unit 1	7
Recommended Reading for Unit 1	7
Broadcast Symbols	7

1.1 Your Pocket Calculator — 8

1.1.1 Introduction — 8
1.1.2 Calculator Arithmetic — 9

1.2 Some Numerical Methods — 14

1.2.1 Non-linear Equations — 14
1.2.2 Simultaneous Linear Equations — 17

1.3 The Approach to Numerical Problems — 23

1.3.1 Choosing the Right Method — 23
1.3.2 Sources of Error — 29

1.4 Flow Charts — 35

1.5 Summary — 41

1.6 Solutions to Self-Assessment Questions — 42

1.7 Glossary and Notation — 49

Guide to the Course

Numerical computation is essentially a practical subject, although it is supported by an ever-increasing body of mathematical theory. Problems which require numerical solutions arise in science, technology, business and economics as well as many other fields. Finding a solution usually involves the following main stages.

(i) Formulation of the problem in mathematical terms. This is often called **modelling**.

(ii) Devising a method of obtaining a numerical solution from the mathematical model.

(iii) Making observations of the numerical quantities which are relevant to the solution of the problem.

(iv) Calculating the solution, usually by means of a computer or at least a hand calculator.

For example, in controlling a space craft on a flight to the moon, the problem is to ensure a safe landing. The mathematical model consists of a set of partial differential equations involving gravitation, the thrust of the rocket engines, the effect of air resistance (when the craft is in the earth's atmosphere), etc. The method used for solving these equations will be numerical in nature; that is to say, it will be a (rather distant) descendant of the Euler method described for ordinary differential equations in *Unit M100 24, Differential Equations I*, rather than of any of the more aesthetically pleasing methods in that unit!

The numerical data which need to be observed will include such things as:

the mass of the craft (which will decrease as the fuel is used up);

the position of the craft relative to the earth and the moon;

the speed and direction of rotation of the craft.

These data will be fed into a computer, which will combine them with the chosen method of solving the equations, and produce solutions consisting of estimates of the corrections required to the craft's course.

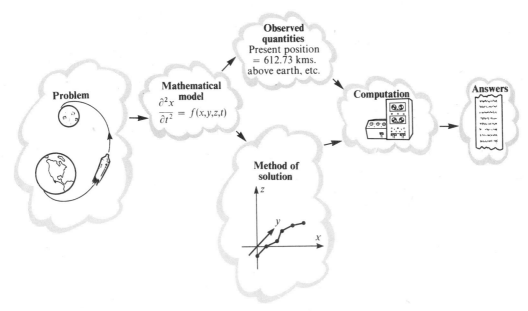

The first of these stages, that of mathematical modelling, is the subject of some other Open University courses, including:

TM281 Modelling by Mathematics (first presentation in 1977),

T341 Systems Modelling.

The second stage, the solution of specific types of mathematical model, is covered by several courses, such as:

M201   Linear Mathematics,

MST282 Mechanics and Applied Calculus,

M321   Partial Differential Equations of Applied Mathematics.

This course concentrates on the second stage (that of devising a method), and also on the fourth stage, that of calculating numerical solutions. However, it is important to bear in mind the whole process of four stages, and to realize that normally what matters is the numerical accuracy of the solution of the overall problem. If it is sufficiently accurate, then the problem solver can claim success. If not, then he has failed, however brilliant his methods may be.

Since a very wide range of calculations is currently carried out by computer, this course has had to be selective in its choice of topics. We have decided to study a fairly small selection in some detail, rather than a wide range more superficially. Our criterion has been to study those methods that occur most frequently in practice, but to omit any that are extensively covered in other Open University courses, unless of particular importance.

One omission of note is that of the numerical solution of differential equations; we found that we were unable to devote enough units to this important subject to do it justice, so rather than skim the surface, we decided to omit the topic altogether. It is in fact covered fairly extensively in M201 (for ordinary differential equations) and in considerable detail in M321 (for partial differential equations).

The course has fifteen units. These are grouped into four blocks, each (except the last) of four units, with each section supported by two television programmes and two radio programmes. In each block the final unit consists of practical work. The television programmes develop topics from the units in the block, and the radio programmes discuss specific key ideas. When listening to the radio programmes, you should have the relevant units with you for reference.

Throughout the course you will find three types of problem in the texts.

(i) Examples

These are problems that are solved as part of the main text.

(ii) Exercises

You should attempt all the exercises and make a real attempt to do them before you turn to the solutions which we have provided. It is often said that mathematics is something which must be learnt by practice. There is probably no branch of mathematics of which this is more true than numerical computation, for this is almost a craft, rather than a science. It relies heavily on experience and intuition, qualities which need to be gradually acquired by practice in solving problems. On the other hand, we do not wish you to exhaust yourself on one individual exercise. Make what is in your view a reasonable attempt and then read the solution, which will be found on the next non-facing page. Whether or not you complete every exercise, always read the solutions. These often contain important ideas which do not appear anywhere else in the course.

(iii) Self-Assessment Questions

These are optional questions for you to practise on, which are intended to help you to monitor your own progress. The solutions are at the end of the unit.

We have tried hard to keep to a minimum the amount of mathematics that we expect you to know before starting the course. Some of the units include a brief summary of any mathematical background assumed, with references to the treatment of the relevant topics.

The Student Computing Service

The practical computing in this course will be carried out by means of the Student Computing Service. Although the art of computer programming is not a major part of the course, we expect you to have used the S.C.S. and written some simple programs in BASIC. If you are uncertain how to use the S.C.S., then you should now read the *Student Computing Service Users Guide*. (If you do not yet possess a copy, you should write to the address at the end of this section.) We will not ask you to write any computer programs, though you may find it useful to do so. Each of the four practical units includes the specification of library programs for the material of the preceding units. Many of these programs will be useful to you before reaching the practical unit, and for this reason we shall try to mail each block of units at approximately the same time.

If your interest in programming goes beyond the requirements of this course, then you would probably like a copy of the *BASIC Reference Manual*. If so, you should write to

> The Open University,
> The Postal Service Supervisor,
> Student Computing Service,
> Walton Hall,
> MILTON KEYNES,
> MK7 6AA.

Study Guide for Units 1 to 4

Units 1 to 4 are a single block of work which forms the material for the Tutor Marked Assignment M351 01. The block consists of the following:

Correspondence Text	Television Programme	Radio Programme
1 Introduction to Numerical Methods	—	1 Errors and Iteration
2 Non-linear Equations	1 Finding Roots of Equations	—
3 Linear Equations	—	2 Matrix Methods
4 Practical Unit I	2 Solving Linear Equations	—

Most of the exercises in Units 1, 2 and 3 can be worked on your pocket calculator, but if you wish to use the Student Computing Service, you will find the corresponding library programs specified in *Unit 4, Practical Unit I*.

Two further television programmes of relevance to this part of the course are:

> *M100 TV 2   Errors that Die,*
>
> *M201 TV 8   Sense and Nonsense with Linear Equations.*

Radio and television broadcasts for this course closely follow the printed texts. There are no separate broadcast notes, but topics covered in the programmes are indicated by notes in the margins of the main texts. The radio broadcasts make specific reference to the texts and you should have your printed units with you while listening to the programmes.

Objectives for Unit 1

We expect that students taking this course will have a wide variety of interests and previous experience. The overall objective of this unit, therefore, is to bring every student up to a level of mathematical and computing ability sufficient to tackle the remaining units of the course. The detailed objectives of the unit are as follows.

(i)   To train you in using the pocket calculator.

(ii) To see how errors arise in numerical calculations and how they propagate.

(iii) To introduce some simple numerical methods which will be developed later in the course. These are:

    (a)   the method of bisection for solving non-linear equations,

    (b)   the Newton–Raphson method for solving non-linear equations,

    (c)   the Gauss elimination method for solving simultaneous linear equations.

(iv) To discuss our approach to numerical problems.

(v)   To introduce the flow chart notation used in this course.

Recommended Reading for Unit 1

The following two books cover much of the material in the early units of the course. Both are well written, reasonably straightforward and full of practical examples.

L. F. Shampine and R. C. Allen, *Numerical Computing: An Introduction* (W. B. Saunders, 1973).

W. S. Dorn and D. D. McCracken, *Numerical Methods with Fortran IV Case Studies* (John Wiley, 1972).

Programming examples in both books are written in FORTRAN, but are kept separate from the main text.

Broadcast Symbols

Throughout the course, topics referred to in Broadcasts are identified by the following symbols.

Radio

Television

The number in the symbol is the programme number.

1.1 YOUR POCKET CALCULATOR

1.1.1 Introduction

You should by now have received your pocket calculator. Full instructions for its use, care and maintenance are given in the *Calculator Handbook*; you should now read this.

It is important to note the **floating point notation** in which the calculator expresses many of its answers. For example, 12 345 678 × 12 345 678 is expressed as

1.5241576	14

meaning

$$1.524\ 157\ 6 \times 10^{14}.$$

Similarly 0.000 123 4 × 0.000 123 4 is expressed as

1.522756	−8

meaning

$$1.522\ 756 \times 10^{-8}.$$

We will go into more detail concerning floating point numbers in *Unit 2*. All we want to mention here is that, though this notation is quite unambiguous on the calculator's display, it is slightly confusing in print. We will therefore adopt a variant of this notation, similar to the computer print-out of very large and very small numbers by the Student Computing Service computer. We will print the calculator displays

1.5241576	14

and

1.522756	−8

as

$$1.524\ 157\ 6 \quad E14$$

and

$$1.522\ 756 \quad E-8$$

respectively.

We will use this notation as and when convenient. For example, we would probably print a number like 0.001 234 567 as it stands, but we would print

$$0.000\ 000\ 000\ 123\ 456\ 7$$

as

$$1.234\ 567 \quad E-10$$

1.1.2 Calculator Arithmetic

The calculator in general records numbers to eight significant figures*. Although this is sufficient for most practical purposes, it means that most numbers cannot be represented exactly. For example, suppose that you use the square root key to calculate the square root of 2. The answer is 1.414 213 5, but if you calculate

$$1.414\ 213\ 5 \times 1.414\ 213\ 5$$

on your calculator, you obtain the answer

$$1.999\ 999\ 8.$$

Moreover, if we try the next largest number that can be represented, we find that

$$1.414\ 213\ 6 \times 1.414\ 213\ 6 = 2.000\ 000\ 1$$

on the calculator. Even if more decimal places were available, we could not find a number which would give exactly 2 when squared. This is not very surprising. It merely illustrates that an irrational number, such as the square root of 2, cannot be expressed exactly as a decimal fraction.

Exercise 1

What are the results of the following calculations on your pocket calculator?

(i) $1 \div 11$

(ii) $(8 \div 9) \times 9$

(iii) $100 \times \pi$ where $\pi = 3.141\ 592\ 653\ 589\ 79\ldots$

These examples show two types of number which cannot be stored as exact decimals in a finite number of digits. They are:

(i)   Irrational numbers, such as π or $\sqrt{2}$, which have no finite representation in any number system.

(ii)   Recurring fractions. Notice that whether or not a number is a recurring fraction depends on the base of the number representation being used. For instance, 1/5 has a simple decimal representation

$$0.2.$$

However, many computers, including those used by the Student Computing Service, represent numbers as binary patterns. In the binary system 1/5 is represented as the recurring fraction

$$0.0011\ 0011\ldots .$$

The fact that not all numbers can be represented exactly in a specified number of digits is of vital importance in computing. The difference between a number and its representation is known as **rounding error**. In the exercise the rounding errors due to inexact arithmetic produced differences of only 1 or 2 in the last of the eight figures. However, in this course we are going to look at computer programs that can run for several hours on large computers which carry out hundreds of thousands of calculations every second. If each of these calculations is wrong by 1 or 2 in the last figure and if all these discrepancies accumulate, then the final result could be so inaccurate as to be of no use whatsoever.

Where a number has more digits than the calculator or computer has room to store, digits will be lost from the right hand side of the number, since these are the least significant digits. The exact effect of this is a little complicated to describe, since there are various ways of doing it, and the result in any case depends on whether the

---

* However, trigonometric, exponential and logarithmic functions are calculated to only six figures.

Solution 1

(i) The exact decimal form of $1 \div 11$ is the recurring decimal $0.0909\ldots$, but the calculator gives $9.0909\ 09\ \ \mathrm{E}-2$, i.e. $9.0909\ 090 \times 10^{-2}$.

(ii) 7.999 999 9.

(iii) Since π cannot be expressed exactly as a decimal number with eight significant figures, we use the approximation

$$\pi = 3.141\ 592\ 6.$$

Then:

$$100 \times \pi = 314.159\ 26.$$

---

machine stores and manipulates the numbers in decimal or binary form. We will return to this question in *Unit 2*, where the binary form of number storage will be described in detail. For the present, it will suffice to say that two sorts of operation can be performed in order to eliminate the extra digits.

(i) Chopping

When a number is **chopped**, the extra digits are simply dropped. Thus, if only four digits can be stored, the (decimal) number 168.692 will be chopped to 168.6, and the (binary) number 101.010 will be chopped to 101.0.

(ii) Rounding

This process in general gives greater accuracy than chopping: the number is **rounded up** or **rounded down** to the nearest number that can be represented by the specified number of digits. Thus, in rounding to four digits, the (decimal) number 168.692 will be rounded up to 168.7 whereas 168.643 will be rounded down to 168.6. In the case of binary numbers, 101.011 will be rounded up to 101.1 and 101.001 will be rounded down to 101.0.

The Student Computing Service computer, a Hewlett Packard 2000F system, works in *binary numbers* and *rounds*; your pocket calculator, on the other hand, works in *decimal numbers* and *chops*. You are asked to confirm this in a self-assessment question.

Note that the errors arising from both rounding and chopping are referred to as rounding errors.

We now want to illustrate some of the problems of computer arithmetic by simple examples on your calculator. In order to keep the examples simple, imagine that you have another calculator that works to four significant digits only, chopping every result. You can use your own calculator to simulate four digit arithmetic if you chop after each stage of the calculation. Thus to calculate

$$3.821 \times (8.646 + 6.332),$$

you first add,

$$8.646 + 6.332 = 14.978,$$

and chop to obtain

14.97.

Now multiply,

$$3.821 \times 14.97 = 57.200\ 37,$$

and chop to obtain the final result,

57.20.

Notice that you must chop after each stage of the calculation.

Exercise 2

Let $a = 76.06$,

and $b = 75.43$.

Both a and b have been obtained by chopping previous results. Calculate the following to four significant figures, chopping after each calculation.

(i) $a + b$

   $a - b$

(ii) $(a + b)(a - b)$

   $a^2 - b^2$

(iii) $(a + b) \div (a - b)$

   $(a - b) \div (a + b)$

State in each case, how many of the digits in your answer are the same as those which would be obtained if this particular step was done by exact arithmetic. In each of the three parts, can you draw any general conclusions about the types of calculation which lead to difficulties?

We can see from this exercise that there are really two possible ways of defining error. The (absolute) **error** in a calculation is the difference between the calculated value and the true value, and will tend to be large if at some stage in the calculation we divide by a very small number (or multiply by a very large number). The **relative error** is the ratio

$$\frac{\text{error}}{\text{number being computed}}$$

and will tend to be large if two nearly equal numbers have been subtracted. It is the relative error which determines the number of significant digits in the computed value which agree with the true value. (These concepts will be defined more formally in *Unit 2*.)

Solution 2

(i)   By calculation to four figures,

$$76.06 + 75.43 = 151.4.$$

All four digits quoted are the same as those which would have been obtained by exact arithmetic.

To four figures,

$$76.06 - 75.43 = 0.6300.$$

This is the same as the solution obtained by exact arithmetic, but the two final zeros are meaningless, since a and b have been obtained by chopping previous results, so that there is no information available about what the last two digits should be in reality. Thus we can only say that two digits are the same as those which would have been obtained by exact arithmetic.

This part demonstrates that if two nearly equal numbers are subtracted, the result may well be accurate to fewer significant digits than either of the original numbers. The same effect will be caused, of course, by adding two numbers of opposite signs but nearly equal magnitudes.

(ii)   To four figures:

$$(a + b)(a - b) = 151.4 \times 0.6300$$
$$= 95.38.$$

By exact arithmetic:

$$(a + b)(a - b) = 151.49 \times 0.63$$
$$= 95.4387.$$

Only the first two digits in the four-figure calculation agree with the calculation by exact arithmetic. The magnitude of the error is 0.0587.

To four figures:

$$a^2 = 5785,$$
$$b^2 = 5689,$$
$$a^2 - b^2 = 96.00.$$

By exact arithmetic:

$$a^2 - b^2 = 95.4387.$$

Only the first figure agrees. The magnitude of the error is 0.5613.

If you compare the results of these two calculations, you will notice that although, when using exact arithmetic,

$$a^2 - b^2 = (a + b)(a - b),$$

this is not necessarily the case when four figure arithmetic is used.

Once again, the inaccuracy is caused by the fact that in each case we have to subtract two nearly equal numbers.

(iii)   To four figures:

$$(a + b) \div (a - b) = 151.4 \div 0.6300$$
$$= 240.3.$$

By exact arithmetic:

$$(a + b) \div (a - b) = 151.49 \div 0.63$$
$$= 240.\dot{4}60\ 31\dot{7}.$$

(The dots above the digits 4 and 7 indicate a recurring decimal.) In this case three digits agree.

The magnitude of the error is 0.160 317 $\dot{4}$.

To four figures:

$$(a - b) \div (a + b) = 0.6300 \div 151.4$$

$$= 0.004\ 161.$$

By exact arithmetic:

$$(a - b) \div (a + b) = 0.63 \div 151.49$$

$$= 0.004\ 158\ 690\ 3\ldots.$$

In this case only two significant digits agree, but the magnitude of the error is only 0.000 002 3.... However, this is quite large relative to the number being computed.

Self-Assessment Questions on Section 1.1

1.   Express the following numbers to five significant digits, both by rounding and by chopping.

   (i)   1.683 962

   (ii) 2487

   (iii) $-4.362\ 18$

   (iv) 496 354.12

2.   Calculate the following, using four-digit arithmetic and chopping.

   (i) $(8.288 + 8.301) - 8.295$

   (ii) $8.288 + (8.301 - 8.295)$

   Which method of calculating $8.288 + 8.301 - 8.295$ would you expect to yield the greater accuracy?

3.   Do the following calculations on your pocket calculator.

   (i) $2 \div 3$

   (ii) $(2 \div 3) - 0.6$

   (iii) $(2 \div 3) \times 3$

   (iv) $1.111\ 111\ 1 + 9.111\ 111\ 8$

   (v) $(-1.111\ 111\ 1) - 9.111\ 111\ 8$

   Determine from these results whether, after calculating a number, the calculator holds any extra digits beyond those appearing on the display, and confirm that it chops after operations of division, multiplication, addition and subtraction.

1.2 SOME NUMERICAL METHODS

1.2.1 Non-linear Equations

A simple problem involving the solution of a non-linear equation, is to calculate a cube root (say, the cube root of 2). Of course, because your pocket calculator possesses a Y^X key, you can do this calculation immediately—or rather, you can pass the buck and trust to the accuracy of the method used by the calculator. However, the blessings of technology are somewhat mixed as regards educative value, and for the purposes of the next exercise we ask you to pretend that your calculator does not have a Y^X key.

Exercise 3

Use your pocket calculator to calculate the cube root of 2 by any method you know or can invent, to as high an accuracy as your calculator will allow.

The solution to this exercise forms the main text for most of this section.

The answer, which is 1.259 921 0, can be found in any one of a variety of ways. The answer itself is not important, but the methods are. Here are three.

(1) Trial and Error

One possible sequence of steps is as follows.

Try 1.5.

$$1.5^3 = 3.375 \qquad \text{(too big).}$$

Try 1.3.

$$1.3^3 = 2.197 \qquad \text{(still too big but fairly close).}$$

Try 1.2.

$$1.2^3 = 1.728 \qquad \text{(too small).}$$

Try 1.27.

$$1.27^3 = 2.048\ 383 \qquad \text{(just too big).}$$

And so on. (Incidentally, you should not be using the Y^X key to calculate cubes!)

A useful approximation to the solution of this problem can probably be found quite quickly by trial and error. However, obtaining the required final result as accurately as the calculator will allow, is likely to be a tedious process. But although trial and error is of little use in obtaining results to a high degree of accuracy, it should not be despised, for it is often useful during the first stages of exploring a problem.

(2) Method of Bisection

Finding the cube root of 2 is equivalent to finding the root of the equation $f(x) = 0$, where

$$f(x) = x^3 - 2.$$

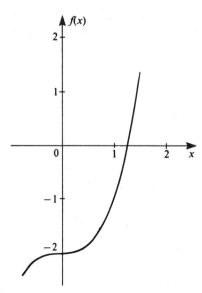

Now $f(1) = -1$,

and $f(2) = +6$.

Since f is a continuous function and the value of f is negative at 1 and positive at 2, then f has (at least) one root in the interval $(1, 2)$*. Most rational human beings would agree that this is obvious; alternatively, it can be proved at great lengths to satisfy a mathematician. (See *Unit M231 4, Three Hard Theorems*!) The midpoint of this interval is 1.5.

$$f(1.5) = (1.5)^3 - 2 = 1.375.$$

Since $f(1)$ is negative, while $f(1.5)$ is positive, we now know that a root lies in the smaller interval $(1, 1.5)$, the midpoint of which is 1.25. We can continue this procedure indefinitely, evaluating f at the midpoint of a sequence of intervals, each being half the size of its predecessor and each known to contain a root since f changes sign somewhere in the interval. Here are the first few calculations.

	Low end of interval	Middle of interval	High end of interval	Value of f at middle
1	1	1.5	2	1.375
2	1	1.25	1.5	−0.046 875
3	1.25	1.375	1.5	0.599 609 3
4	1.25	1.3125	1.375	0.260 986 2
23	1.259 920 9	1.259 921 0	1.259 921 1	−0.000 000 3

Thus after 23 bisections and evaluations of f, we obtain, to eight figures:

$$\sqrt[3]{2} = 1.259\ 921\ 0.$$

(3)   Newton–Raphson Method

If you took M100, you should have studied this method in *Unit M100 14, Sequences and Limits II*. The method is also to be found in *Unit MST281 10, Taylor Approximation*, but it is in an optional section, and therefore if you took MST281 you may possibly not be familiar with it. We will therefore give a brief description of it at this point; those of you who are confidently familiar with the technique may like to proceed straight to Exercise 4.

The Newton–Raphson method is one of a number of **iterative methods** of solving equations of the form $f(x) = 0$. The fundamental feature of an iterative method is that an approximation is made to a root of an equation, and a sequence of successively better approximations is then generated. Thus, if x_0 is our first approximation to a root of $f(x) = 0$, our iterative method prescribes a formula for getting from x_0 to x_1 (a hopefully better approximation); from x_1, the same formula then gives us x_2; and we eventually produce a sequence

$$x_0, x_1, x_2, \ldots ,$$

which (we hope!) converges to a root of $f(x) = 0$. (See *Unit M100 7, Sequences and Limits I*, or *Unit MST281 2, Functions and Limits*.)

The Newton–Raphson method relies on the fact that, given a function which can be written down in terms of some formula, it is often possible to write down a formula for its derivative. We can therefore find a formula for the tangent to the curve at any given point, and a tangent to a reasonably smooth curve is usually a good approximation to the curve itself (within a certain region).

---

* We use square brackets $[a, b]$ to denote $\{x : a \leq x \leq b\}$, and round brackets (a, b) to denote $\{x : a < x < b\}$.

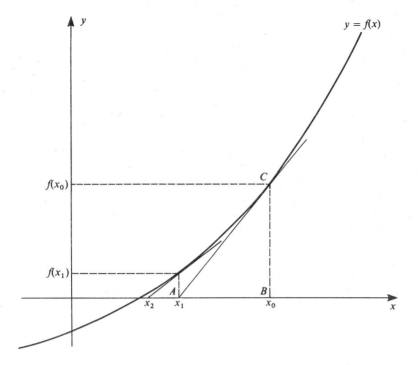

In the above diagram, for example, if x_0 is our first approximation to the root of $f(x) = 0$, then clearly x_1 is a much better approximation, and x_2 is very much better still. Apparently, therefore, the sequence

$$x_0, x_1, x_2, \ldots$$

converges quite rapidly to the root of $f(x) = 0$. (We have not *proved* this, of course; but we will go into the details of the method and its justification in *Unit 2*.)

So, how do we calculate x_1 from x_0? Look at the right-angled triangle ABC in the figure; the gradient of f at x_0 is the ratio of the sides BC and AB. This leads to the equation

$$f'(x_0) = \frac{f(x_0)}{x_0 - x_1},$$

i.e.

$$x_1 = x_0 - \frac{f(x_0)}{f'(x_0)}.$$

Applying the same formula repeatedly, we calculate x_{n+1} from x_n on the $(n + 1)$th step:

$$x_{n+1} = x_n - \frac{f(x_n)}{f'(x_n)}.$$

This is the **Newton–Raphson formula**.

For the function

$$f(x) = x^3 - 2$$

we have the particularly simple form

$$x_{n+1} = \frac{2}{3}\left(x_n + \frac{1}{x_n^2}\right).$$

16

Using this expression, with a first estimate of 1, we get the following sequence of estimates on our calculator.

n	x_n
0	1
1	1.333 333 3
2	1.263 888 8
3	1.259 933 4
4	1.259 921 0
5	1.259 921 0

Since $x_5 = x_4$, all further estimates will have the same value, and this is our final answer.

Exercise 4

Use the Newton–Raphson method to calculate the cube root of 2000 on your calculator. Is it sensible to set $x_0 = 1$ in this case?

Self-Assessment Questions on Section 1.2.1

4. (i) Show that there is exactly one positive root of the equation

$$\frac{x}{2} = \sin x$$

where the angle is measured in radians.

(ii) Verify that the root is in the interval $[1, 2]$, and use the method of bisection twice to obtain a smaller interval in which the root must lie.

(iii) Use the Newton–Raphson method to determine the root as accurately as your calculator will allow.

5. An analogue of the method of bisection can be used to find a local maximum of a continuous function f. If $a < b < c$, and $f(a) < f(b) > f(c)$, then f has at least one local maximum between a and c. Use this method to find a local maximum point of the function

$$f(x) = 1 - x^2 - \tfrac{1}{2}e^{-x},$$

to one significant figure.

(Take $[0, 0.5]$ as your first interval.)

1.2.2 Simultaneous Linear Equations

Many problems can be reduced to solving a set of simultaneous linear equations. This is the topic of *Unit 3*, T.V. Programme 2 and some of *Practical Unit I*. In this section we want to introduce the fundamental method of solution, Gauss elimination, and to illustrate that, under unfavourable circumstances, round-off errors can build up very rapidly. So that we can illustrate the build up of errors in quite small problems, we use four digit arithmetic, with truncation after each stage of calculation.

Here are two sets of three equations in which each coefficient is expressed to four figures.

(i) $0.3333x_1 + 1.000x_2 + 0.7500x_3 = 2.600$

$\phantom{0.3333x_1 + {}} 7.474x_2 + 4.184 \ x_3 = 15.06$

$\phantom{0.3333x_1 + 7.474x_2 + {}} 1.495 \ x_3 = 1.230$

17

Solution 4

In this case, the recurrence formula is:

$$x_{n+1} = \frac{2}{3}\left(x_n + \frac{1000}{x_n^2}\right).$$

Letting $x_0 = 1$, the first few iterations give:

n	x_n
0	1
1	667.333 33
2	444.890 36
3	296.596 94
4	197.738 86
5	131.842 95

For $n > 1$, each x_n is approximately $\frac{2}{3}$ of its successor at this stage of the iteration. Since the final answer is in the region of 12, this is clearly most inefficient. A better choice of initial estimate method would be $x_0 = 10$. We then get the following iteration.

n	x_n
0	10
1	13.333 333
2	12.638 888
3	12.599 334
4	12.599 210
5	12.599 210

This exercise highlights the need to choose a sensible first approximation when using an iterative method; a bad first approximation will often lead to an enormously increased amount of work, even when the correct solution is finally obtained.

(ii) $0.3333x_1 + 1.000\ x_2 + 0.7500x_3 = 2.600$

$-2.162\ x_1 + 0.9888x_2 - 0.6800x_3 = -1.800$

$1.000\ x_1 - 2.500\ x_2 + 0.6667x_3 = -2.050$

Exercise 5

Find x_1, x_2 and x_3 which satisfy the first set of equations. Use four digit arithmetic throughout, chopping at each stage.

The method of solution can be summarized as follows.

(i)   The final equation is solved to find the value of the final variable; this value is substituted in the other equations.

(ii) The last unsolved equation is now solved to find the value of the next variable.

(iii) The process is repeated until the values of all the variables have been found.

This method is known as **back substitution**. This first set of equations was easy to solve, because the matrix of coefficients on the left-hand side of the equation was in **upper triangular form**:

$$\begin{bmatrix} * & * & * \\ 0 & * & * \\ 0 & 0 & * \end{bmatrix}.$$

Note that all the numbers below the main diagonal (that is, top left to bottom right) are zeros. Any set of linear equations in upper triangular form can be solved by back substitution as in the solution to this exercise, as long as there are no zeros on the main diagonal.

A set of n simultaneous equations in n unknowns which is not in upper triangular form can be replaced by an equivalent set of equations in upper triangular form by means of a process called **Gauss elimination**. By **equivalent** we mean that the two sets of equations have the same solution. (We shall temporarily ignore any errors that are introduced by using inexact arithmetic.)

Gauss elimination uses the idea of an **elementary operation** on an equation. In any set of equations, the equality will still hold if any of the following operations is carried out.

(i)   Multiply both sides of an equation by a non-zero constant.

(ii) Add a multiple of one equation to another.

(iii) Interchange two equations.

These operations provide simple methods for solving simultaneous linear equations. In fact in its simplest form, Gauss elimination uses only the second type of elementary operation. To demonstrate how it works, we shall solve the second set of equations given at the beginning of this section. We adopt the convention of writing only the coefficients of the equations and leaving out the x_1, x_2, etc. Thus the equations are as follows.

Left-hand side

Right-hand side

$$\begin{bmatrix} 0.3333 & 1.000 & 0.7500 \\ -2.162 & 0.9888 & -0.6800 \\ 1.000 & -2.500 & 0.6667 \end{bmatrix} \qquad \begin{bmatrix} 2.600 \\ -1.800 \\ -2.050 \end{bmatrix}$$

We want to reduce this set of equations to upper triangular form by using the second type of elementary operation. Suppose we add m times the first equation to the second, giving

$$(0.3333m - 2.162)x_1 + (1.000m + 0.9888)x_2 + (0.7500m - 0.6800)x_3$$
$$= (2.600m - 1.800).$$

This will produce a zero in the first position in the new equation if

$$0.3333m - 2.162 = 0,$$

from which we obtain

$$m \simeq 6.486 \qquad \text{(four digit arithmetic)}.$$

Solution 5

From the third equation,

$$x_3 = 1.230/1.495$$

$$= 0.8227.$$

The second equation can now be solved for x_2.

$$7.474x_2 + 4.184 \times 0.8227 = 15.06,$$

$$x_2 = 1.553.$$

The top equation can now be solved for x_1.

$$0.3333x_1 + 1.000 \times 1.553 + 0.7500 \times 0.8227 = 2.600,$$

$$x_1 = 1.290.$$

Thus the solution that we have computed is

$$x_1 = 1.290$$

$$x_2 = 1.553$$

$$x_3 = 0.8227.$$

Using this approximate value of m, the new second equation has coefficients.

Left-hand side	Right-hand side

0	7.474	4.184	15.06.

In a similar way we can get a zero in the first position of the third equation by adding n times the first equation to the third equation, where n is chosen so that

$$0.3333n + 1.000 = 0,$$

from which we obtain

$$n = -3.000.$$

Thus the original set of equations is replaced by the following set.

Left-hand side	Right-hand side

$$\begin{bmatrix} \mathbf{0.3333} & 1.000 & 0.7500 \\ 0 & 7.474 & 4.184 \\ 0 & -5.500 & -1.583 \end{bmatrix} \qquad \begin{bmatrix} 2.600 \\ 15.06 \\ -9.85 \end{bmatrix}$$

During each stage of the manipulations the current element in the main diagonal is known as the **pivot**; the row and column in which the pivot lies are called the **pivot row** and the **pivot column**. For this first stage of the elimination, the pivot is the number 0.3333, which we have printed in bold type.

The above matrix of coefficients is not yet in upper triangular form, because there is still a non-zero element in the second position of the final equation. This can be eliminated by adding a suitable multiple of the second equation to the third equation, that is, by repeating the previous stage using the last two equations only.

Exercise 6

Reduce the last set of equations to upper triangular form by using the second type of elementary operation.

(i)   What is the pivot?

(ii) What is the multiplier?

(iii) What is the new set of equations?

The main steps of the Gauss elimination method are as follows.

Step 1.   Select the first row as the pivot row.

Step 2.   Add a suitable multiple of the pivot row to the next row so as to obtain a zero below the pivot.

Step 3.   Repeat step 2 for every row below the pivot row.

Step 4.   Repeat steps 2 and 3 for each successive pivot.

Note that every element of the main diagonal, except the last, becomes the pivot in its turn.

Although our written summary of Gauss elimination is rather awkward, the method itself seems very simple. Here apparently is a straightforward method for reducing any number of simultaneous linear equations to upper triangular form so that they can then be solved by back substitution. However, before becoming too confident, let us try the same method on the following apparently innocuous set of equations.

$$1.62x_1 + 1.10x_2 + 0.65x_3 = 3.37$$

$$6.18x_1 + 4.20x_2 - 3.04x_3 = 7.34$$

$$4.65x_1 - 3.05x_2 + 2.10x_3 = 3.70.$$

At a quick glance these equations seem perfectly harmless, and since the coefficients are expressed to three significant figures, one might well expect that four digit arithmetic would be sufficiently accurate.

Using the Gauss elimination method, with four digit arithmetic, we go through the following stages. First we add suitable multiples of the first row to each subsequent row in turn, to obtain zeros in the first column. (We will drop the labelling of the left and right hand sides.)

$$\begin{bmatrix} \mathbf{1.62} & 1.10 & 0.65 & 3.37 \\ 0 & 0.005 & -5.519 & -5.51 \\ 0 & -6.207 & 0.235 & -5.971 \end{bmatrix}$$

Then we add a suitable multiple of the second row to the third to reduce the set of equations to upper triangular form.

$$\begin{bmatrix} 1.62 & 1.10 & 0.65 & 3.37 \\ 0 & \mathbf{0.005} & -5.519 & -5.51 \\ 0 & 0 & -6848 & -6842 \end{bmatrix}$$

Back substitution now gives the following values.

$$x_3 = 0.9991$$

$$x_2 = 0.8000$$

$$x_1 = 1.135.$$

You can easily check that the original set of equations have in fact the solution $x_1 = x_2 = x_3 = 1$. Thus our calculated values of x_1 and x_2, in particular, contain serious errors.

Solution 6

(i)   In this case, the pivot is 7.474.

(ii) We add p times the second equation to the third, choosing p so that the new equation has a zero in the second position:

$$7.474p - 5.500 = 0,$$

which gives the multiplier

$$p \simeq 0.7358.$$

(iii) The new third equation is

Left-hand side	Right-hand side
0 0 1.495	1.23.

We have now reduced the original set of equations to the upper triangular form:

Left-hand side	Right-hand side

$$\begin{bmatrix} 0.3333 & 1.000 & 0.7500 \\ 0 & 7.474 & 4.184 \\ 0 & 0 & 1.495 \end{bmatrix} \qquad \begin{bmatrix} 2.600 \\ 15.06 \\ 1.23 \end{bmatrix} .$$

This is in fact the first set of equations which was given at the beginning of this section, and which you have already solved by back substitution.

If you work through this calculation yourself, you can probably discover where the errors arose and grew. The coefficient 0.005, which is the second pivot, was obtained by subtracting two nearly equal numbers one of which involves a small rounding error. Although the error in this term is small in absolute size, it is a relatively large proportion of the coefficient. In the second stage of the elimination, this pivot was used to calculate a multiplier, by dividing into -6.207. As we have seen, division by a small number multiplies any errors. Thus a small rounding error has been magnified to produce major errors in the final result. At present we leave the phenomenon as a curiosity, but in *Unit 3* we shall return to Gauss elimination, find out what has gone wrong and look at some methods of avoiding such a situation.

Self-Assessment Question on Section 1.2.2

6. (i)   Solve the following set of simultaneous linear equations using Gauss elimination and back substitution. Notice that this set is exactly the same as the previous example except that they are arranged in a different order. Use four digit arithmetic.

$$1.62x_1 + 1.10x_2 + 0.65x_3 = 3.37$$

$$4.65x_1 - 3.05x_2 + 2.10x_3 = 3.70$$

$$6.18x_1 + 4.20x_2 - 3.04x_3 = 7.34.$$

  (ii) What does this suggest to you about Gauss elimination and back substitution?

  (iii) Is there an even better order in which to place the three equations, reducing still further the possible errors in the solution?

1.3 THE APPROACH TO NUMERICAL PROBLEMS

1.3.1 Choosing the Right Method

The task of the numerical problem solver is to find answers with high numerical accuracy, rather than to construct pleasing mathematical generalizations. Thus, his method of approaching a particular problem may well differ considerably from that which a pure mathematician would instinctively adopt. The following example illustrates this quite graphically.

Example

A man with a rifle having a muzzle velocity of 626.0 metres per second stands on a raised platform with the end of the muzzle 15.00 m vertically above the ground (which is level), and aims at a point on the ground 20.00 m away in a horizontal direction. All the above numbers are correct to four significant figures. The problem is to find (also to four figures) where the bullet strikes the ground. Suppose that you have a calculator which works to four figures (or, alternatively, a set of four-figure log. tables).

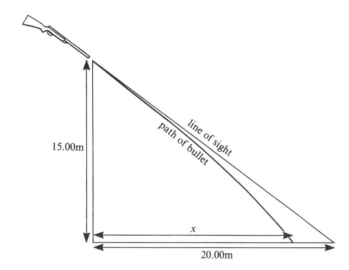

Intuition tells us that, in such a short distance, the bullet will deviate by very little from a straight line. (The deviation is, of course, magnified in the figure.) Thus, quite a good approximation would be

$$x = 20.00 \text{ m}.$$

For a more accurate answer, we must take gravity into account. Here we run across an apparent difficulty: g, the acceleration due to gravity, is normally given as 9.81 metres per second per second, correct to only three significant figures. This is because g varies slightly over the earth's surface, so that the fourth and higher figures are meaningless. As we are trying to calculate to four significant figures, this seems to pose a problem. However, gravity will cause only a slight deviation in the bullet's flight in this particular case, so if we can calculate the deviation to three figures, this will probably allow us to find x to four figures.

The pure mathematician might proceed rather along the following lines. This is a particular case of a general situation in which the rifle makes an angle of α degrees with the horizontal, is h metres above the ground, and has muzzle velocity v. Take a Cartesian co-ordinate system whose origin is on the ground directly below the muzzle of the rifle, with x the measure of horizontal distance and y the measure of vertical distance.

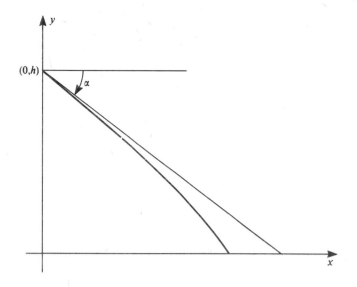

In the most general case, the rifle might be pointing at an upward or a downward angle. It would be natural to regard α as positive for an upward angle and negative for a downward angle, so in our particular case,

$$\alpha = -\tan^{-1}\left(\frac{15.00}{20.00}\right).$$

If t represents the time that has elapsed since firing the bullet, then we have two equations:

$$x = vt \cos \alpha \qquad (1)$$

$$y = h + vt \sin \alpha - \tfrac{1}{2}gt^2 \qquad (2)$$

The problem is to find x when $y = 0$. Expressing t in terms of x by means of equation (1):

$$t = \frac{x}{v \cos \alpha}.$$

Substituting this expression into equation (2) and setting $y = 0$, we arrive at the following quadratic equation for x.

$$\left(\frac{-g}{2v^2 \cos^2\alpha}\right)x^2 + x \tan \alpha + h = 0. \qquad (3)$$

Let $a = \dfrac{-g}{2v^2 \cos^2\alpha}$, $b = \tan \alpha$, and $c = h$. Then we want the positive solution of equation (3). Since a is negative, this is:

$$x = \frac{-b - \sqrt{b^2 - 4ac}}{2a}. \qquad (4)$$

We now find that, if we use four-figure arithmetic to evaluate x, we obtain embarrassingly different results depending on whether we round or chop at each stage!

(i) Rounding

We obtain:

$$a = -1.956 \quad \text{E}-5$$

$$b = -0.7500$$

$$c = 15.00.$$

Thus:

$$x = \frac{0.7500 - \sqrt{0.5625 + 0.0011\ 74}}{-3.912\ \ \text{E}-5}.$$

Now the square root of 0.5637 is 0.7508, rounding to four figures, and therefore the next stage in the calculation gives

$$x = \frac{-8 \quad E-4}{-3.912 \quad E-5}$$

and there is only one significant figure left in the numerator! If we nevertheless press on regardless, we obtain our final answer:

$$x = 20.45.$$

This implies that the bullet curves upwards rather than downwards, which is absurd.

(ii) Chopping

Despite the fact that chopping is theoretically less accurate than rounding, we get a result this time which is at least on the right side of ridiculous. We obtain:

$$a = -1.954 \quad E-5$$

$$b = -0.7500$$

$$c = 15.00.$$

The next stage gives

$$x = \frac{0.7500 - 0.7507}{-3.908 \quad E-5}$$

$$= \frac{-7 \quad E-4}{-3.908 \quad E-5}$$

$$= 17.91.$$

This implies that the bullet falls more than two metres short, which does not sound very accurate, even though it avoids the absurdity of the answer obtained by rounding!

The reason for the failure is not too far to seek: it lies in the subtraction of the two very nearly equal numbers $-b$ and $\sqrt{b^2 - 4ac}$. There are several ways open to us of tackling this problem. Perhaps the simplest is to proceed as follows. (For the sake of argument, let us use rounding rather than chopping.)

Clearly, x is close to 20.00. Therefore, since a is very small, ax^2 is very close to $a \times (20.00^2)$; in other words, the absolute difference between ax^2 and $-400.0 \times 1.956 \quad E-5$ is extremely small. Therefore a very good approximation to the quadratic equation (3) is the linear equation:

$$-400 \times 1.956 \quad E-5 + x \tan \alpha + h = 0,$$

i.e.

$$-0.007\ 824 - 0.75x + 15.00 = 0,$$

which has the solution to four figures:

$$x = 19.99.$$

In effect, we have used an iterative method to solve this equation. We have re-expressed the equation

$$ax^2 + bx + c = 0$$

as

$$x = g(x)$$

where

$$g(x) = \frac{-c - ax^2}{b},$$

and used the iteration formula

$$x_{n+1} = g(x_n).$$

In saying that x was close to 20.00, we were in effect choosing

$$x_0 = 20.00.$$

After one iteration,

$$x_1 = 19.99,$$

and in fact another iteration gives

$$x_2 = 19.99.$$

This strongly indicates that we have found the solution to four figures.

Thus: a (mathematically) approximate method of solving this particular equation gives a (numerically) far more accurate answer than the (mathematically) exact method, and with far less work!

There is quite an important point to note here. Suppose you used the formula

$$x = \frac{-b - \sqrt{b^2 - 4ac}}{2a}$$

and calculated everything using the full accuracy of the calculator, rather than just four figures. You would obtain

$$x = 19.989\ 400 \text{ metres},$$

which agrees well with our iterative method. Thus, using a large number of significant figures can overcome the problem of a numerically inappropriate method. Occasionally, there may be problems where all known methods of solution have associated numerical problems; in this case, the only cure is to work to as many significant figures as possible. However, this is a counsel of despair. One should always look first for a numerical method that does not involve a build-up of errors.

Assuming you do have a satisfactory method, using the calculator, it will often involve calculating several intermediate results and writing these down for further use. Clearly, it is a waste of effort to write these numbers down to the full eight-figure accuracy if the data has only, say, three significant figures. On the other hand, it is wise to have one or two extra significant figures in reserve. Exactly how far to go in this respect really depends on the problem; numerical analysis is something of an art as well as a science. But as a general rule of thumb, if you are confident that there are no serious numerical problems in your method, we would give the following advice.

If the crucial pieces of data are given to n figures, then it is *usually* safe to record intermediate calculations to $n + 2$ figures.

Thus, in the gun example, the crucial pieces of data are to four figures. (Gravity is only given to three figures, but its effect on the answer is comparatively small.) Therefore, if you were using your calculator in the normal way rather than going through the exercise of chopping to four figures, you would record a, b and c to six figures.

Exercise 7

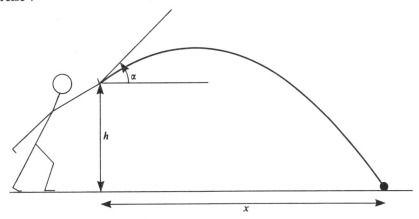

The figure shows an athlete putting the shot. The problem is to find how far the shot is thrown. The equation to be solved is the same as that in the example we have just discussed. To three significant figures, you may take in this case:

$g = 9.81$ metres per second per second,

$\alpha = 45.0$ degrees,

$h = 2.00$ metres,

$v = 9.90$ metres per second.

(i)   Set up the quadratic equation for x in the form

$$ax^2 + bx + c = 0.$$

(ii) Does the formula

$$x = \frac{-b - \sqrt{b^2 - 4ac}}{2a}$$

provide a numerically satisfactory method of solving the problem to three significant figures?

(iii) Try to solve the problem by the Newton–Raphson method, taking $x_0 = 0$ as the initial estimate. What goes wrong?

Solution 7

(i) From the example, the general equation is:

$$\frac{-gx^2}{2v^2\cos^2\alpha} + x\tan\alpha + h = 0.$$

This is of the form $ax^2 + bx + c = 0$. Calculating a, b and c to five figures as our rule of thumb suggests:

$$a = -0.100\ 09,$$

$$b = 1,$$

$$c = 2.$$

(ii) The formula gives:

$$x = \frac{-1 - \sqrt{1.800\ 72}}{-0.2}$$

$$= 11.710.$$

There is no subtraction of nearly equal numbers here, and we can be reasonably confident that the solution is 11.7 metres correct to three significant figures.

(iii) We have:

$$f(x) = -0.100\ 09x^2 + x + 2,$$

$$f'(x) = -0.200\ 18x + 1.$$

The recurrence formula gives, after simplifying:

$$x_{n+1} = \frac{0.100\ 09x_n^2 + 2}{0.200\ 18x_n - 1}.$$

Starting with $x_0 = 0$, we get the following table.

n	x_n
0	0
1	-2
2	$-1.714\ 102\ 0$
3	$-1.708\ 010\ 9$
4	$-1.708\ 008\ 1$

The iteration certainly converges, but the answer is clearly not the one required. We have found the wrong root of the equation; if the graph of the trajectory is projected backwards, we can see what has happened.

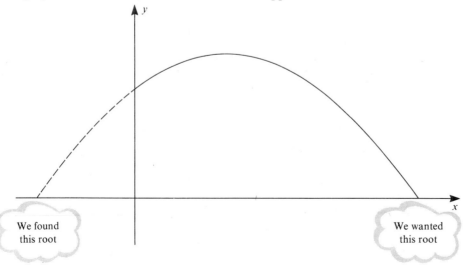

We found this root

We wanted this root

Exercise 7, part (iii) was intended to give you some insight into the Newton–Raphson method for solving non-linear equations, and in particular to emphasize the importance of choosing a suitable starting value. In *Unit 2* we will examine the method of bisection and the Newton–Raphson method in more detail.

1.3.2 Sources of Error

In section 1.0 we drew a diagram of the essential steps involved in producing a solution to a numerical problem. Stripped of its florid detail, the essence of the diagram is as follows.

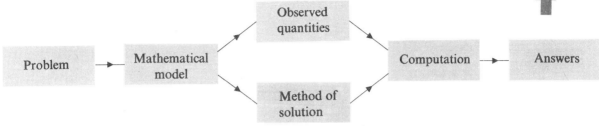

There are several stages in the process where errors can creep in. This can be illustrated by making the following additions to the diagram.

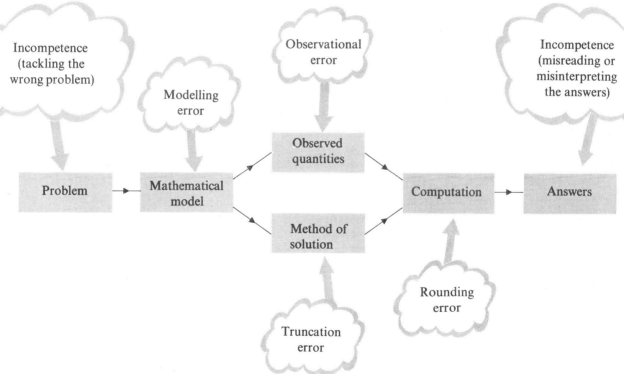

We shall ignore the possibility of error due to incompetence! Thus, there are four unavoidable sources of error.

(1) Modelling error

This usually occurs because the mathematical model does not give an exact description of the problem. Various factors are frequently omitted, either because it would be too complicated to take them into account (e.g. the effect of air resistance in the example of the previous section), or because the effects are simply not known, and various simplifying assumptions must be made. Very often, modelling errors occur because it is assumed that a system being studied can in effect be isolated from outside influences. Thus, in sending a space craft to the moon, the gravitational effect of the earth, moon and sun would probably be taken into account, but not that of any of the other planets. The assumption would be made that, for the purposes of navigation,

the system consisting of craft, earth, moon and sun was isolated from the rest of the universe. Similarly, in the example of the previous section, the bullet and the solid earth are considered in isolation from the air molecules. In this case, the modelling error would make very little difference to the accuracy of the result in the particular example given, but could result in great inaccuracy if we were, say, firing the bullet in an upward rather than a downward direction.

(2)   Observational error

There are two sorts of numerical data, namely those arrived at by:

(i)   counting,

(ii)   measuring.

The first type of data comprise such things as:

the number of cards in a pack, if one is calculating the probabilities of various poker hands;

the number of *yes*, *no* and *don't know* responses if one is conducting a public opinion poll.

The second type are exemplified by:

the height of the gun above the ground in the example of the previous section;

the velocity of the shot in Exercise 7.

In principle, error need not occur at all in the first type of data (though in practice it often will). However, observational error is always present in data arrived at by measurement, since in this case the quantity being measured can vary over a continuous range of values, and every method of observing any such quantity has only a certain limited accuracy. Sometimes, attempts to specify a quantity to more than a certain accuracy are simply meaningless. For instance, the muzzle of a rifle has a certain diameter (of the order of 1 cm, say), and so in our gun example there would be no point at all in trying to specify the position of the end of the muzzle to an accuracy of 0.01 cm.

(3)   Truncation error

As you proceed through this course, you will see a large number of numerical methods of solving mathematical problems. The essence of any numerical method is that it is approximate. The approximation usually occurs because of **truncation**; this is rather similar to the chopping of a number, except that it is the *method* that is chopped. For example, in an iterative method for solving an equation, the solution is theoretically obtained as the limit of a process involving an infinite number of steps, but in practice the method is truncated so that the solution is quoted after only a finite number of steps. Similarly, in an equation involving (for example) $\cos \theta$, where θ is known to be small, we might expand $\cos \theta$ in a Taylor series:

$$\cos \theta = 1 - \frac{\theta^2}{2!} + \frac{\theta^4}{4!} \cdots$$

and then truncate to get an approximate expression for $\cos \theta$:

$$\cos \theta \simeq 1 - \frac{\theta^2}{2!}.$$

A theoretically infinite series is truncated to a finite series.

(4)   Rounding error

Rounding error occurs at every stage in the calculation by the computer (or calculator). In most calculations, the vast majority of rounding errors will consist of the chopping or rounding that occurs after an arithmetic operation. However, some rounding errors in the computer are more like truncation errors: they occur when the computer is asked to evaluate certain functions. For example, suppose a line in a program reads:

110   LET Y = SIN(X)

Then the computer will evaluate sin x by some approximate method (such as taking a finite number of terms from a series).

In fact, the chopping of a number can also be regarded as a form of truncation. A general number is the sum of an infinite series of powers of 10 (in decimal) or 2 (in binary), and chopping corresponds to taking only a finite number of these terms. Thus, from a mathematical point of view, truncation and rounding errors are very similar. But from a practical point of view they are quite different. Truncation errors are under the control of the person using the computer to solve his problem. If he thinks that the truncation error arising out of some method is too large, he can change the method, such as by taking more terms of a series. On the other hand, rounding errors are a fundamental property of any computer system, and are not under the control of the user.

Of course, the user can exert a considerable influence on the *effect* of rounding error on his final answer. The example of the previous section is a good instance of this. Using an exact method with no truncation error (the formula for the roots of a quadratic) led to a huge rounding error. The preferable strategy was to choose a method with a small truncation error, but which led the computer to commit only a small amount of rounding error. The combination of truncation and rounding error using this method was far less than the rounding error alone using the more naive method.

In this course, we shall largely neglect modelling and observational error, and concentrate on truncation and rounding error, and on the art of choosing a method of solution which minimizes these. Our example in the previous subsection is well worth revisiting from time to time, as it illustrates in a mathematically simple context, the philosophy and methodology of the whole course.

Exercise 8

(i)   You are standing on a cliff, looking across a strait 48.0 kilometres wide. You can just see the top of a cliff on the other side of the water. Assuming that your eye level is 60.0 metres above the water, how high is the cliff across the water? To four significant figures, the earth's mean radius is 6367 kilometres.

   You may find the following diagram useful.

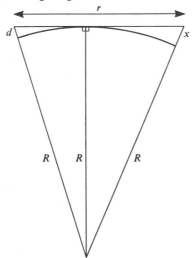

(ii)   Identify the various forms of error involved in the process of arriving at your answer.

(iii) How badly is your answer affected if you assume that there is a slight swell on the sea, giving waves 1 metre high? (That is, the crests are 1 metre above mean sea level; the vertical distance from crest to trough is 2 metres.)

31

Solution 8

(i) In the diagram, x is the height of the cliff across the water, d is the height of your eye above sea level, r is the width of the strait, and R is the earth's radius. Thus r is the sum of two distances: from the eye to the point where the line of sight touches the sea, and from that point to the top of the other cliff. Pythagoras' Theorem applied to the two right-angled triangles in the diagram gives:

$$r = \sqrt{(R + x)^2 - R^2} + \sqrt{(R + d)^2 - R^2}$$
$$= \sqrt{2Rx + x^2} + \sqrt{2Rd + d^2}.$$

(Note that, by expressing $(R + x)^2 - R^2$ and $(R + d)^2 - R^2$ as $2Rx + x^2$ and $2Rd + d^2$ respectively, we are eliminating a situation where, at first sight, we appear to have to subtract pairs of nearly equal numbers.)

To obtain a quadratic equation in x (the only unknown), put the term containing x on one side, then square both sides.

$$x^2 + 2Rx = [r - \sqrt{2Rd + d^2}]^2.$$

Substituting the given values, we obtain (with the distances expressed in kilometres):

$$x^2 + 12\ 734x = 414.48.$$

(Again, we are using the rule of thumb that it is only worthwhile recording two more significant figures than the least accurate of the data. In fact, d is so small compared with R that it makes no difference, to five significant figures, whether the d^2 term is taken into account or not.)

As in the gun example, the positive root of this equation is small, so the x^2 term is small compared with the others, and an efficient and accurate method is the iterative method:

$$x_0 = \frac{414.48}{12\ 734},$$

$$x_1 = \frac{414.48 - x_0^2}{12\ 734},$$

etc.

We obtain:

$$x_0 = 0.032\ 549\ 081,$$

$$x_1 = 0.032\ 548\ 998,$$

$$x_2 = 0.032\ 548\ 998.$$

Since some of the original data is given to only three significant figures, it is reasonable to express the answer as:

$$x = 32.5 \text{ metres.}$$

(ii) The iteration method has introduced a very small truncation error; by rounding some of our intermediate results to five figures we have introduced small rounding errors. But our observational errors are considerably larger, since the height of the observer above sea level, and the width of the strait, are given to only three figures. Finally, there are several modelling errors:

(a) We have assumed that the earth is a perfect sphere; this is quite a small error.

(b) We have assumed that light travels in straight lines. This may sound like a reasonable assumption, but in fact if the air close to the sea has a temperature gradient, light passing through it will undergo a very slight change in direction. Since the distance it has to travel (48 kilometres) is very large compared with the two heights involved, this possibility could introduce serious errors. (Mirages are examples of this effect, and their potentially catastrophic consequences are well known!)

(c) It is not clear precisely what is meant by saying that you can just see the top of the cliff. What you will detect is a breaking of the otherwise smooth horizon. By how much does the horizon have to be disturbed in order for your optic nerve to trigger a recognition response in your brain? This will depend on factors such as how clear the air is, and whether the cliff top is bare and smooth, or whether on the other hand it has distinctive shapes such as buildings on it.

(d) We have neglected the influence of waves on the sea.

(e) We have neglected tides.

(iii) We now have to assume that the line of sight just skims the top of the waves.

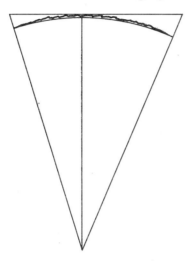

The simplest way to express this assumption is to measure the observer's height and the height of the cliffs from the tops of the waves. This means that the observer must now consider himself to be 59.0 metres above the sea. Working through the calculation and replacing 60.0 by 59.0 in the data, the new answer becomes (to three figures):

33.3 metres.

This is the height from the tops of the waves. Therefore the height of the cliff above sea level is

34.3 metres.

Thus, the effect of quite small waves means that our answer is trustworthy to only one significant figure. (It is probable that the other modelling errors which we mentioned have an even worse effect. All we can really say, after all this calculation, is that the cliff is almost certainly at least 30 metres high, more probably 40 or 50!)

Self-Assessment Questions on Section 1.3

7. In the gun example of section 1.3.1, there is in fact a non-iterative method of solving equation (3) which is numerically accurate. This is based on the fact that, for the general quadratic equation

$$ax^2 + bx + c = 0,$$

the product of the two roots is $\dfrac{c}{a}$. Now the formula

$$x = \frac{-b + \sqrt{b^2 - 4ac}}{2a}$$

gives a numerically accurate method of finding the negative root of equation (3). Thus the positive root is

$$x = \left(\frac{c}{a}\right) \div \left(\frac{-b + \sqrt{b^2 - 4ac}}{2a}\right)$$

$$= \frac{2c}{-b + \sqrt{b^2 - 4ac}}.$$

Solve the problem by this method, using four-figure arithmetic and rounding at each stage. Compare the number of arithmetic operations necessary with the number required by the iterative method.

8. Yet another variation on the example of section 1.3.1. This time, you are standing above a stretch of level ground, and you cast a heavy stone down towards the ground. As it leaves your hands, it has a velocity of 0.1 metres per second at a downward angle of 60.0 degrees to the horizontal. If your hands are 20.0 metres above the ground, what horizontal distance does the stone travel? (Try to find a convenient and accurate iterative method.)

9. A sphere of radius 1 unit has a segment cut from it by a plane; the volume of the segment is 1/100 of the volume of the sphere. Find as accurately as your calculator will allow, the thickness of the segment, i.e. the distance d in the figure.

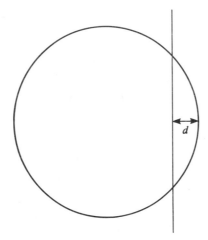

To do this problem, you will need the formula for the volume of a solid of revolution, which you will have met in *Unit M100 9* or *Unit MST281 8*. To save you looking it up, the volume of the solid formed by rotating the non-negative function $y = f(x)$, between $x = a$ and $x = b$, about the x-axis, is

$$\pi \int_a^b (f(x))^2 \, dx.$$

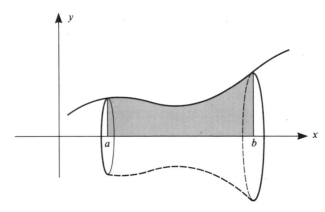

1.4   FLOW CHARTS

We assume at the start of this course that you have used a computer before at least to some extent. Which computer is not significant, but it is important that you should have had at least a little experience of drawing flow charts and writing computer programs in BASIC. We shall use the following notation for our flow charts. It is not the same in every respect as the notation which is used in other Open University courses.

Input

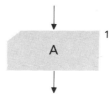

We use a box shaped like a punched card. The figure shows an instruction to input a value into a location called A.

Output

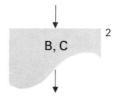

The shape of this box represents the paper (torn across the bottom) that emerges from a line printer. This figure shows an instruction to print the values in B and C.

Assignment

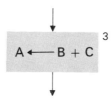

Assignment instructions to change the current values of variables are written in rectangular boxes. The figure shows an instruction to assign to A the sum of the values in the locations of B and C. The arrow points backwards to indicate that the value of B + C is put back into variable A, and to remind you that it is *not a function arrow* of the sort met in M100/MST 281.

Decisions

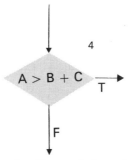

Decision instructions are written in diamond-shaped boxes. Our decisions will usually be taken by determining whether a statement is true (abbreviated T) or false (F). Thus, the figure shows a decision whether the value in A is greater than the sum of the values in B and C.

Start and Stop

To be complete, a flow chart must contain both a START and a STOP box.

Iteration Boxes

Many computer programs involve a repetitive operation where the number of times that it has to be repeated is specified by a counter. Loops controlled by a counter occur so frequently and are so similar in their structure that it is convenient to use a special notation for them, which allows us to compress three boxes into one. Consider the following fragment of a flow chart.

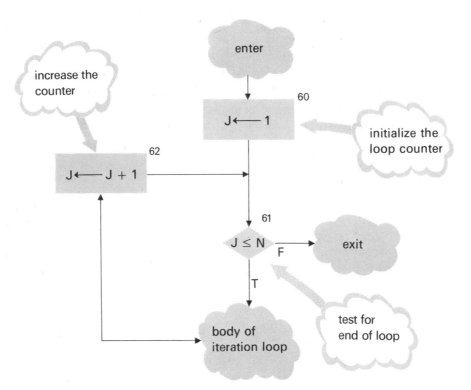

The effect of boxes 60, 61 and 62 is to ensure that the loop is passed through N times. In our more compact notation the steps numbered 60, 61 and 62 in the flow chart fragment are written in just one box in this way.

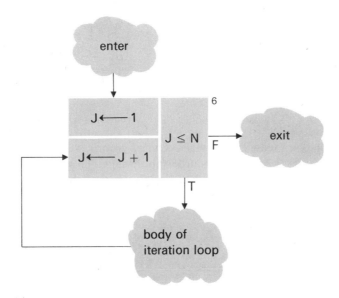

This notation may well seem strange to you at first, but it should soon become familiar and easy to use.

Arrays

An **array** is a simple way of representing related items of data by the use of subscripts. An array using one subscript is also called a **list** or a **vector**.

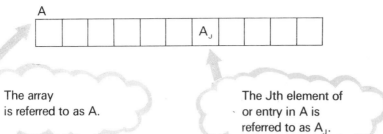

The array is referred to as A.

The Jth element of or entry in A is referred to as A_J.

If two subscripts are used the array may also be called a **table** or a **matrix**.

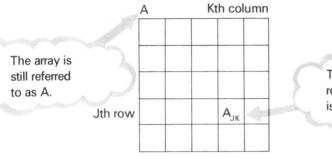

The array is still referred to as A.

The entry in the Jth row and Kth column is referred to as A_{JK}.

We shall use the notation

$$\{A_J : J = 1, 20\}$$

to stand for the first 20 entries in the array A.

Thus the following box

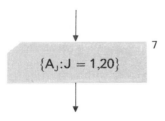

$$\{A_J : J = 1, 20\}$$

is an instruction to input twenty values into the vector A in the order $A_1, A_2, \ldots, A_{20}$.

Any other flow chart notation that we adopt will be described when it is first used in the course.

To illustrate the use of our flow chart notation, we now draw a flow chart of the Newton–Raphson method for finding the root of a function given an initial guess for the root and a tolerance to be applied to the function.

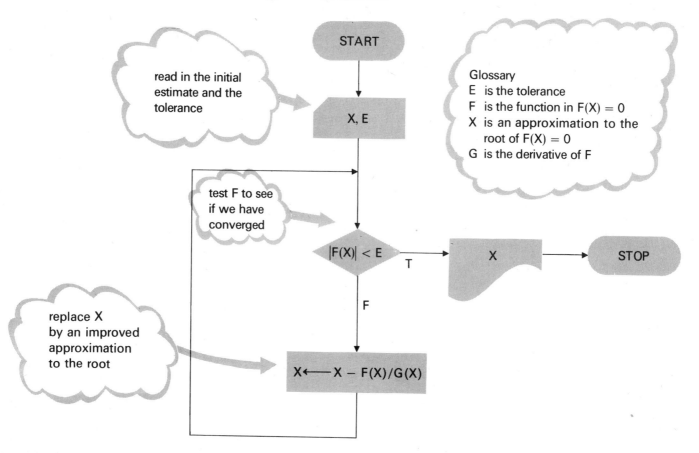

This is obviously a very simplified version of the method which will give good results for many problems. However, there are difficulties which may arise which the program will not be able to deal with.

(i)   The initial estimate X may cause the iterative process to diverge or converge to the wrong root.

(ii) If E is too small then we will never get out of the loop.

(iii) If E is too large then we may leave the loop too early without getting an answer sufficiently close to the root.

(iv) The method will break down completely if G(X) = 0 at any stage in the iteration process.

These difficulties will be investigated in *Unit 2*.

Exercise 9

Draw a flow chart for calculating the matrix product AV where A is a matrix and V is a suitable column vector (i.e. one-column matrix), and printing out the result.

Self-Assessment Question on Section 1.4

10. Draw a flow chart of the method of bisection used to solve a non-linear equation, assuming that we know a pair of values A and B such that F(A) and F(B) have opposite signs. You may use the fact that there is a SGN function in BASIC such that

$$SGN(X) = \quad 1 \quad \text{if } X > 0$$
$$= \quad 0 \quad \text{if } X = 0$$
$$= -1 \quad \text{if } X < 0.$$

Solution 9

This flow chart will illustrate the use of **nested loops** to carry out a complex arithmetical operation.

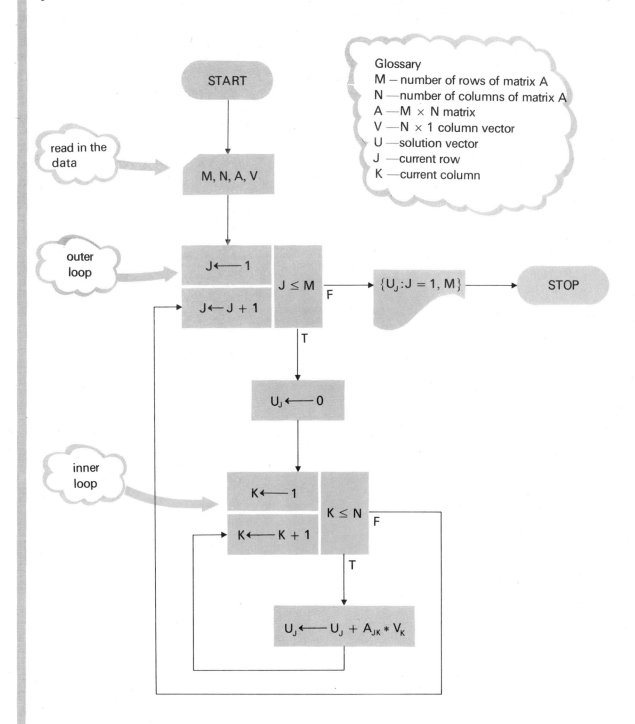

The first loop of the nest specifies the row of the matrix A, while the second loop denotes the process of multiplying the vector V by the Jth row of the matrix.

1.5 SUMMARY

(i)   Calculators and computers store the results of calculations to a given number of significant digits; a number with too many significant digits to be stored thus is **chopped** or **rounded**.

(ii) The **absolute error** in a calculation will tend to become large if at some stage we divide by a very small number or multiply by a very large number. The **relative error**, on the other hand, is likely to become large if two nearly equal numbers are subtracted. It is the relative error which determines the number of significant digits in the calculation which agree with the true value.

(iii) Two simple methods for solving a non-linear equation are the **method of bisection** and the **Newton–Raphson method**.

(iv) Simultaneous linear equations can be solved by using **Gauss elimination** to reduce the matrix of coefficients to **upper triangular form**. The solution can then be found by **back substitution**. Under unfavourable circumstances **rounding errors** can build up rapidly with the basic form of this method.

(v)   The four main sources of error in a numerical calculation are: **modelling error**, **observational error**, **truncation error** and **rounding error**. In this course, we concentrate on the last two of these. They are not independent of each other; often, a theoretically exact method of solution (with no truncation error) may yield large rounding errors, which can be substantially reduced by choosing a different method, albeit one involving some truncation error.

(vi) **Flow charts** are a useful technique for describing numerical algorithms, especially if a computer is to be used in the solution of the problem.

1.6 SOLUTIONS TO SELF-ASSESSMENT QUESTIONS

1. *Rounding*   *Chopping*

    (i)   1.6840   1.6839

    (ii) Cannot be done, as the original number has only four significant digits.

    (iii) −4.3622   −4.3621

    (iv) 496 350   496 350

2. (i)   8.288 + 8.301 = 16.58;

          16.58 − 8.295 = 8.285.

    (ii) 8.301 − 8.295 = 0.006;

          0.006 + 8.288 = 8.294.

    Since method (ii) involves no chopping, it must yield the correct answer if the initial numbers are exact, and is likely to give the greater accuracy in the final answer if they are inexact.

3. (i) $2 \div 3 = 6.666\ 666\ 6\ \ E-1$.

    (ii) $(2 \div 3) - 0.6 = 6.666\ 666\ \ E-2$.

    (iii) $(2 \div 3) \times 3 = 1.999\ 999\ 9$.

    (iv) $1.111\ 111\ 1 + 9.111\ 111\ 8 = 10.222\ 222$.

    (v) $(-1.111\ 111\ 1) - 9.111\ 111\ 8 = -10.222\ 222$.

    If the calculation held any extra digits beyond those appearing on the display, then the result of (ii) would have eight 6's, not seven. Thus, the calculator does not hold any extra digits after a number has been calculated.

    The result of (i) shows that the calculator chops after division; otherwise, the answer would be $6.666\ 666\ 7\ \ E-1$.

    Consider the result of (iii). The calculation

    $$6.666\ 666\ 6\ \ E-1 \times 3$$

    gives the nine-digit number

    $$1.999\ 999\ 98.$$

    If this were then rounded to eight digits, the answer would be 2. But it is actually 1.999 999 9, therefore the calculator chops after multiplication.

    Consider the result of (iv). The calculation gives the nine-digit number

    $$10.222\ 222\ 9.$$

    To get the eight-digit answer 10.222 222, the calculator therefore chops after addition.

    Similarly, the result of (v) shows that the calculator chops after subtraction.

4. (i)   Consider the graphs of $y = \dfrac{x}{2}$ and $y = \sin x$.

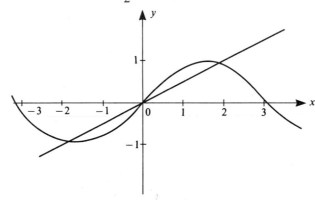

It is clear that there are three solutions of $\dfrac{x}{2} = \sin x$; one negative, one zero and one positive. A more rigorous argument would be as follows.

Consider the function

$$f(x) = \sin x - \frac{x}{2}.$$

Then

$$f'(x) = \cos x - \tfrac{1}{2},$$
$$f''(x) = -\sin x.$$

Now $f(0) = 0$. Between 0 and $\cos^{-1}(\tfrac{1}{2})$, f is a strictly increasing function, since $f'(x) > 0$, therefore there are no solutions in $(0, \cos^{-1}(\tfrac{1}{2})]$. Between $\cos^{-1}(\tfrac{1}{2})$ and π, f is a strictly decreasing function, since $f'(x) < 0$, therefore there is *at most* one solution in $(\cos^{-1}(\tfrac{1}{2}), \pi]$. (If α, β were solutions, we would have $f(\alpha) = f(\beta) = 0$, contradicting the fact that f is strictly decreasing in this interval.) Now $f(\cos^{-1}(\tfrac{1}{2})) > 0$, and $f(\pi) = -\dfrac{\pi}{2} < 0$, therefore there is a solution in this interval.

Finally, for $x > \pi$, we have $f(x) < \sin x - \dfrac{\pi}{2} < 0$, as $|\sin x| \le 1$ and $\dfrac{\pi}{2} > 1$.

(ii) Using the calculator:

$$f(1) = \cdot \ 3.414\ 71 \quad \text{E}-1,$$
$$f(2) = -9.070\ 3 \quad \text{E}-2.$$

Thus, there is a root of $f(x) = 0$ in $[1, 2]$. Bisecting $[1, 2]$:

$$f(1.5) = 2.474\ 95 \quad \text{E}-1.$$

Bisecting $[1.5, 2]$:

$$f(1.75) = 1.089\ 86 \quad \text{E}-1.$$

(iii) It is evident from part (ii) that

$$x_0 = 2$$

would be a good initial estimate for the Newton–Raphson method. The formula gives:

$$x_{n+1} = x_n - \frac{\sin x_n - \tfrac{1}{2}x_n}{\cos x_n - \tfrac{1}{2}}.$$

This gives the following sequence on the calculator.

n	x_n
0	2
1	1.900 995 2
2	1.895 511 8
3	1.895 494 9
4	1.895 494 4
5	1.895 494 2
6	1.895 494 1
7	1.895 494 1

Thus the answer is

$$x = 1.895\ 494\ 1.$$

5. $f(0) = 0.5,$

$f(0.5) = 4.467\ 345\ \ E-1.$

Bisecting the interval:

$f(0.25) = 5.480\ 995\ \ E-1.$

Therefore $f(0) < f(0.25) > f(0.5)$, and there is a local maximum in $[0, 0.5]$.

We now bisect both the intervals $[0, 0.25]$ and $[0.25, 0.5]$:

$f(0.125) = 5.431\ 265\ \ E-1.$

$f(0.375) = 5.157\ 305\ \ E-1.$

Therefore $f(0.125) < f(0.25) > f(0.375)$, and there is a local maximum in $[0.125, 0.375]$.

Continuing in this way:

$f(0.1875) = 5.503\ 293\ \ E-1,$

$f(0.3125) = 5.365\ 358\ \ E-1,$

local maximum in $[0.125, 0.25]$.

$f(0.156\ 25) = 5.479\ 135\ \ E-1,$

$f(0.218\ 75) = 5.503\ 870\ \ E-1,$

local maximum in $[0.1875, 0.25]$.

Thus, to one significant figure, there is a local maximum at 0.2.

6. (i) Subtract suitable multiples of the first row from the other rows.

$$\begin{bmatrix} \mathbf{1.62} & 1.10 & 0.65 & \bigm| & 3.37 \\ 0 & -6.207 & 0.235 & \bigm| & -5.971 \\ 0 & 0.005 & -5.519 & \bigm| & -5.51 \end{bmatrix}$$

Subtract a suitable multiple of the second row from the third row.

$$\begin{bmatrix} 1.62 & 1.10 & 0.65 & \bigm| & 3.37 \\ 0 & \mathbf{-6.207} & 0.235 & \bigm| & -5.971 \\ 0 & 0 & -5.518 & \bigm| & -5.514 \end{bmatrix}$$

Back substitution gives the following.

$x_3 = 0.9992$

$x_2 = 0.9996$

$x_1 = 1.000.$

(ii) This shows that the accuracy of the Gauss elimination method can *sometimes* be improved by changing the order of the equations in the system being solved.

(iii) If the equation with the largest coefficient of x_1 is written first, then we will multiply it by numbers less than 1 when we subtract it from the other rows, so that there will be still less magnification of the error at this stage.

$$\begin{bmatrix} 6.18 & 4.20 & -3.04 & \bigm| & 7.34 \\ 1.62 & 1.10 & 0.65 & \bigm| & 3.37 \\ 4.65 & -3.05 & 2.10 & \bigm| & 3.70 \end{bmatrix}$$

On the second step, we can if necessary rearrange the second and third equations so that once again we multiply the relevant row by a number less than 1 before subtracting it. This method of minimizing the error by rearranging the equation is known as **partial pivoting**, and is discussed in detail in *Unit 3*.

7. Using four-figure arithmetic and rounding, we go through the following steps.

$$x = \frac{30.00}{0.7500 + \sqrt{0.5637}} \quad \text{(four multiplications, one addition)}$$

$$= \frac{30.00}{0.7500 + 0.7508} \quad \text{(one square root operation)}$$

$$= 19.99 \quad \text{(one addition, one division)}.$$

This gives a total of four multiplications, two additions, one square root operation and one division, after the process of calculating a, b and c (which is the same for both methods).

On the other hand, the iterative method estimates x_0 as $\dfrac{-c}{b}$ (one division), then calculates x_1 from x_0 by two multiplications, a subtraction and a division. Since x_1 is as accurate as possible for four figures, this gives a total of two multiplications, two divisions and a subtraction. The iterative method still wins on computational efficiency, by a short head!

8. This is the same equation as in the example in section 1.3.1, with the following data:

$g = 9.81$ metres per second per second,

$\alpha = -60.0$ degrees,

$h = 20.0$ metres,

$v = 0.1$ metres per second.

Then the equation to be solved is

$$ax^2 + bx + c = 0,$$

where (working to five figures):

$$a = \frac{-g}{2v^2 \cos^2\alpha} = -1962.0,$$

$$b = \tan\alpha \qquad = -1.7321,$$

$$c = h \qquad\quad = 20.0.$$

In this case, the formula

$$x = \frac{-b - \sqrt{b^2 - 4ac}}{2a}$$

involves no subtraction of nearly equal numbers, and gives a reliable result. However, the question asks for a convenient iterative method. In this case, the x term is much smaller than the x^2 term, so to a first approximation we have:

$$x_0^2 = \frac{-c}{a};$$

$$x_0 = \sqrt{-\frac{c}{a}}.$$

This leads to the iteration:

$$x_{n+1}^2 = \frac{-c - bx_n}{a},$$

$$x_{n+1} = \sqrt{\frac{-c - bx_n}{a}},$$

and we obtain

$$x_0 = 1.009\ 637\ 4\quad E-1,$$

$$x_1 = 1.005\ 213\ 7\quad E-1,$$

$$x_2 = 1.005\ 233\ 1\quad E-1,$$

$$x_3 = x_4 = 1.005\ 233\ 0\quad E-1.$$

Thus, to three figures,

$$x = 0.101 \text{ metres.}$$

9. Taking the centre of the sphere as origin, the sphere is the solid of revolution generated by a unit circle, with equation

$$x^2 + y^2 = 1,$$

i.e.

$$y = \sqrt{1 - x^2}.$$

Suppose a plane perpendicular to the x axis cuts the sphere as follows.

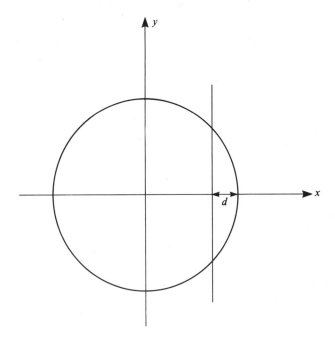

Then by the formula for the volume of a solid of revolution, the volume of the segment cut from the sphere is

$$\pi \int_{1-d}^{1} y^2\,dx$$

$$= \pi \int_{1-d}^{1} (1 - x^2)\,dx$$

$$= \pi(d^2 - \tfrac{1}{3}d^3).$$

Now the volume of the whole sphere is

$$\pi \int_{-1}^{1} (1 - x^2)\,dx = \tfrac{4}{3}\pi.$$

Therefore the equation to be solved is:

$$d^2 - \tfrac{1}{3}d^3 = \tfrac{1}{100} \times \tfrac{4}{3}$$

$$= \tfrac{1}{75};$$

$$d^3 - 3d^2 + \tfrac{1}{25} = 0.$$

46

Since d is rather small compared with 1, d^3 will be rather small compared with d^2, and one possible iteration formula would be:

$$d_{n+1}^2 = \tfrac{1}{3}(d_n^3 + \tfrac{1}{25}),$$

i.e.
$$d_{n+1} = \sqrt{\tfrac{1}{3}(d_n^3 + \tfrac{1}{25})},$$

$$d_0 = \sqrt{\tfrac{1}{75}} = 1.154\ 700\ 4 \quad \text{E} - 1,$$

This gives, after finding that $d_5 = d_6$:

$$d = 1.178\ 062\ 6 \quad \text{E} - 1.$$

However, the ratio of d^2 to d^3 is not all that great—it is of the order of 10 to 1. Thus, the Newton–Raphson method is likely to require considerably fewer iterations. This gives the formula:

$$d_{n+1} = d_n - \frac{d_n^3 - 3d_n^2 + \tfrac{1}{25}}{3d_n^2 - 6d_n}$$

$$= d_n - \frac{d_n^2(d_n - 3) + 0.04}{3d_n(d_n - 2)}.$$

This gives almost the same answer as above after finding that $d_2 = d_3$:

$$d = 1.178\ 062\ 8 \quad \text{E} - 1^*.$$

10. We have to have as input:

(i) the function f whose root we are to find,

(ii) values a and b such that $f(a)$ and $f(b)$ are of opposite sign;

(iii) a value ε for the tolerance, i.e. the size of the interval within which the root is to be located. (For example, if the root is to be located to 4 significant figures, and we have found an interval $[x, x + \varepsilon]$ where $\varepsilon = 0.0001$, within which the root must lie, then $x + \dfrac{\varepsilon}{2}$ is an acceptable estimate of the root.)

There are two possible reasons for stopping the program:

(i) the size of the interval is less than the tolerance, so that we have found a satisfactory estimate for the root,

(ii) an end of one of the intervals (say x) is such that $f(x)$ is calculated to be exactly zero, in which case x is the best estimate we can give for the root.

(We have not taken incompetence into account: if the initial values a and b do not give opposite sign, then we may be in trouble. But the following flow chart does not take account of this.)

Bearing these points in mind, the following is one possible flow chart (but by no means the only possible one).

* This is without using the Y^x key. As this key gives only 6-digit accuracy, you will have obtained a slightly different result if you used it.

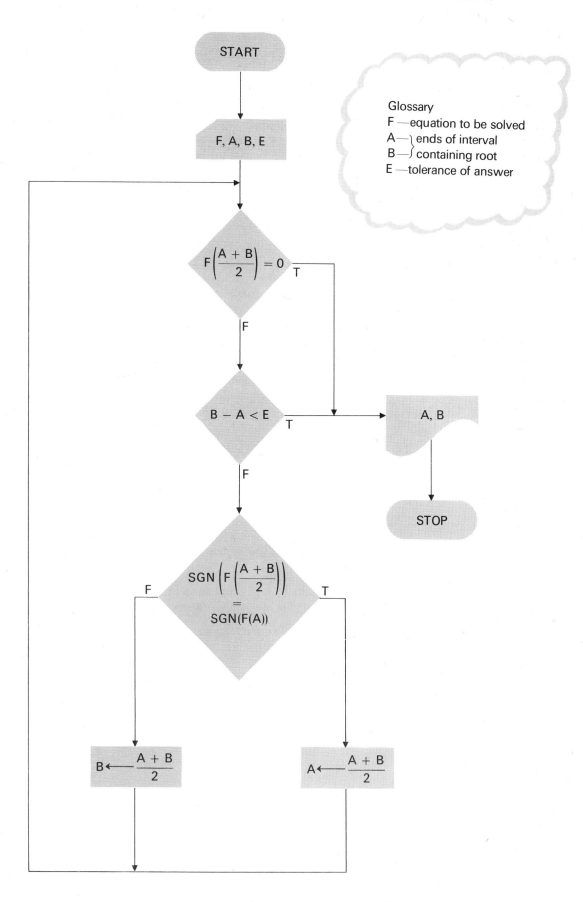

START

F, A, B, E

Glossary
F —equation to be solved
A —⎫ ends of interval
B —⎭ containing root
E —tolerance of answer

$F\left(\dfrac{A+B}{2}\right) = 0$ T

F

$B - A < E$ T

A, B

STOP

$SGN\left(F\left(\dfrac{A+B}{2}\right)\right)$
$=$
$SGN(F(A))$

F   T

$B \leftarrow \dfrac{A+B}{2}$

$A \leftarrow \dfrac{A+B}{2}$

1.7 GLOSSARY AND NOTATION

Glossary

array	page 37
assignment box	page 35
back substitution	page 19
chop	page 10
decision box	page 35
elementary operation	page 19
equivalent (sets of equations)	page 19
error (absolute)	page 11
floating point notation	page 8
flow chart	page 35
Gauss elimination	page 19
input box	page 35
iteration box	page 36
iterative method	page 15
list	page 37
matrix	page 37
method of bisection	page 14
modelling	page 4
modelling error	page 29
nested loops	page 40
Newton–Raphson formula	page 16
Newton–Raphson method	page 15
observational error	page 30
output box	page 35
pivot (column, row)	page 20
relative error	page 11
round (down, up)	page 10
rounding error	pages 9, 30
START box, STOP box	page 35
table	page 37
truncation error	page 30
upper triangular form	page 19
vector	page 37

Notation

1.524 157 6 E14	page 8
flow chart notation	pages 35 to 37

NUMERICAL COMPUTATION COURSE UNITS

1 Introduction to Numerical Methods
2 Non-linear Equations
3 Linear Equations
4 Practical Unit I
5 Linear Programming I
6 Linear Programming II
7 Integer Programming
8 Practical Unit II
9 Approximation I
10 Approximation II
11 Integration
12 Practical Unit III
13 Simulation I
14 Simulation II
15 Practical Unit IV
16 *no text*